Design Supplement
Engineering Mechanics
Statics

Seventh Edition

Design Supplement Engineering Mechanics Statics

Seventh Edition

Russell C. Hibbeler

PRENTICE HALL Upper Saddle River, NJ 07458

Prentice-Hall, Inc.
A Simon & Schuster/Viacom Company
Upper Saddle River, NJ 07458

10 9 8 7 6 5 4 3 2 1

ISBN 0-13-567959-1

Printed in the United States of America

Design Supplement Engineering Mechanics Statics

Seventh Edition

[To be assigned after coverage of Section 6.2 or 6.4]

D–1. DESIGN OF A BRIDGE TRUSS

A bridge having a horizontal top cord is to span between the two piers *A* and *B*. It is required that a pin-connected truss be used, consisting of steel members bolted together to steel gusset plates, such as the one shown in the figure. The end supports are assumed to be a pin at *A* and a roller at *B*. A vertical loading of 5 kN is to be supported within the middle 3 m of the span. This load can be applied in part to several joints on the top cord within this region, or to a single joint at the middle of the top cord. The force of the wind and the weight of the members are to be neglected.

Assume the maximum tensile force in each member cannot exceed 4 kN; and regardless of the length of the member, the maximum compressive force cannot exceed 2.5 kN. Design the most economical truss that will support the loading. The members cost $3.50/m, and the gusset plates cost $8.00 each. Submit your cost analysis for the materials, along with the scaled drawing of the truss, identifying on this drawing the tensile and compressive force in each member. Also, include your calculations of the complete force analysis.

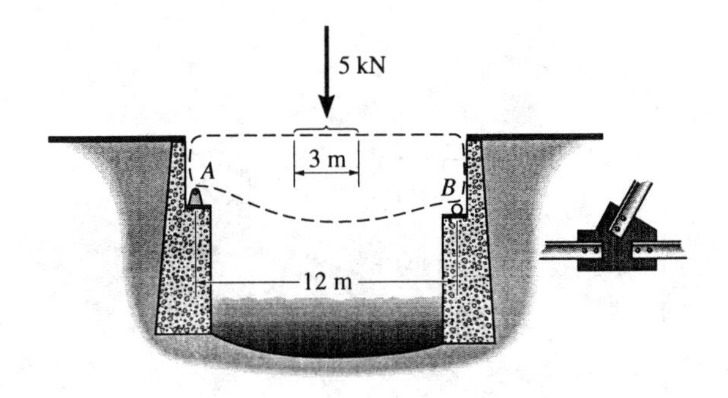

D–2. DESIGN OF A ROOF TRUSS

A roof is to be suspended between two walls. It is required that a pin-connected truss be used, consisting of wood members nailed together with steel gusset plates, such as the one shown in the figure. The end supports are assumed to be a pin at A and a roller at B. A vertical loading of 1500 lb is to be supported within the middle 15 ft of the span. This load can be applied in part to several joints along the bottom cord within this region, or to a single joint. Neglect the force of wind on the roof and the weight of the members.

The height h of the truss is restricted to be within the range 3 ft $\leq h \leq$ 15 ft. Assume the maximum tensile force in each member cannot exceed 750 lb; and regardless of the length of the member, the maximum compressive force cannot exceed 300 lb. Design the most economical truss that will support the loading. The wood members cost \$0.75/ft, and the gusset plates cost \$2.50 each. Submit your cost analysis for the materials, along with a scaled drawing of the truss, identifying on this drawing the tensile and compressive force in each member. Also, include your calculations of the complete force analysis.

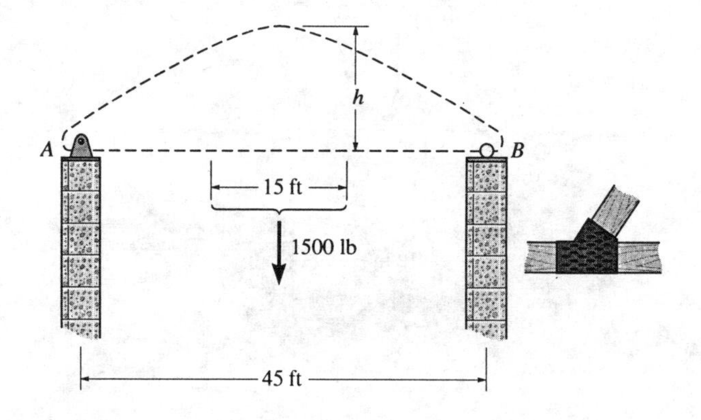

[To be assigned after coverage of Section 6.6]

D–3. DESIGN OF A PULLEY SYSTEM

The steel beam *AB,* having a length of 5 m and a mass of 7 Mg, is to be hoisted in its horizontal position to a height of 4 m. Design a pulley-and-rope system, which can be suspended from the overhead beam *CD,* that will allow a single worker to hoist the beam. Assume that the maximum (comfortable) force that he can apply to the rope is 180 N. Submit a drawing of your design, specify its approximate material cost, and discuss the safety aspects of its operation. Rope costs $1.25/m and each pulley costs $3.00.

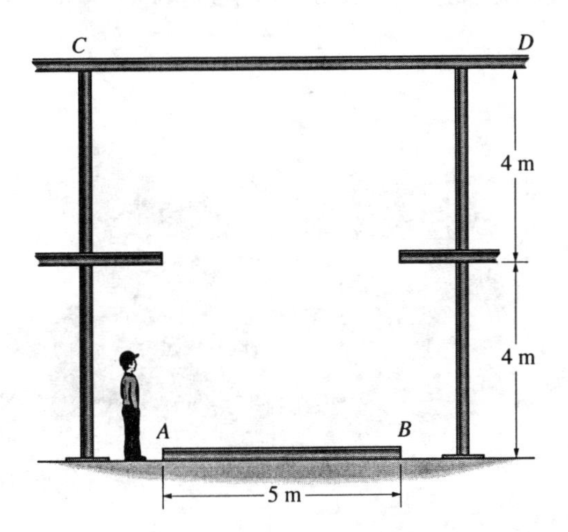

D–4. DESIGN OF A DEVICE TO LIFT CONCRETE BLOCKS

Concrete blocks used for erosion control of bridge and highway embankments are manufactured with a locking groove in them having the dimensions shown. If each block has a weight of 500 lb, design a device that can be used to lift a block by fitting the device down within the groove. (Slipping the device into the ends of the groove is prohibitive, since the blocks are stacked and shipped next to each other.) The device is to be made of a smooth material. Submit a scaled drawing of your device, along with a brief explanation of how it works, based on a force analysis. Also, discuss the safety aspects for its use.

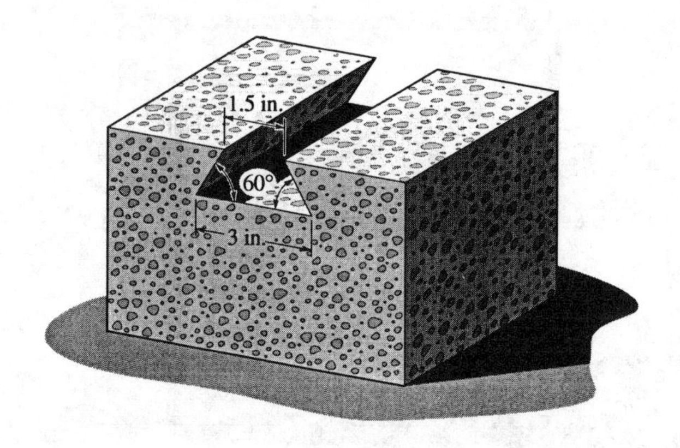

D–5. DESIGN OF A TOOL USED TO POSITION A SUSPENDED LOAD

Heavy loads are suspended from an overhead pulley and each load must be positioned over a depository. Design a tool that can be used to shorten or lengthen the pulley cord *AB* a small amount in order to make the location adjustment. Assume the worker can apply a maximum (comfortable) force of 25 lb to the tool, and the maximum force in cord *AB* is 500 lb. Submit a scaled drawing of the tool, and a brief paragraph to explain how it works using a force analysis. Include a discussion on the safety aspects of its use.

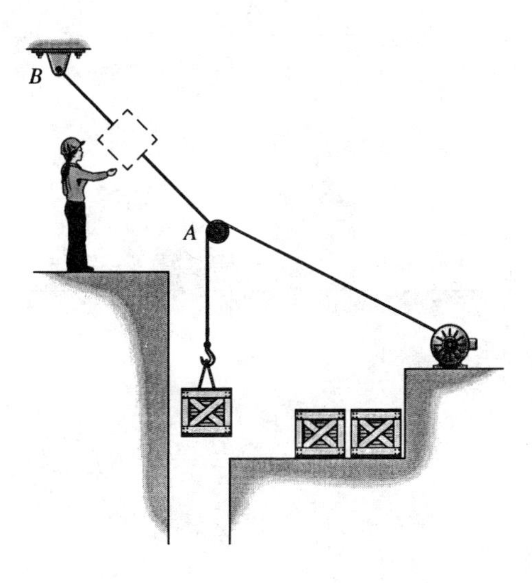

D–6. DESIGN OF A CAN CRUSHER

In an effort to conserve resources a homeowner decides to make a device to crush standard soft-drink cans for recycling. Design a can-crushing device that operates by hand with an applied force of 3 lb, and is easy to construct and operate. Submit a scale drawing of the device, including a force analysis, and discuss the safety and reliability of its use. For this project, perform an experiment to obtain the magnitude of force needed to crush an empty can to a height of 1 in.

D–7. DESIGN OF A FENCE-POST REMOVER

A farmer wishes to remove several fence posts. Each post is buried 18 in. in the ground and will require a maximum vertical pulling force of 175 lb to remove it. He can use his truck to develop the force, but he needs to devise a method for their removal without breaking the posts. Design a method that can be used, considering the only materials available are a strong rope and several pieces of wood having various sizes and lengths. Submit a sketch of your design and discuss the safety and reliability of its use. Also, provide a force analysis to show how it works and why it will cause minimal damage to a post while it is being removed.

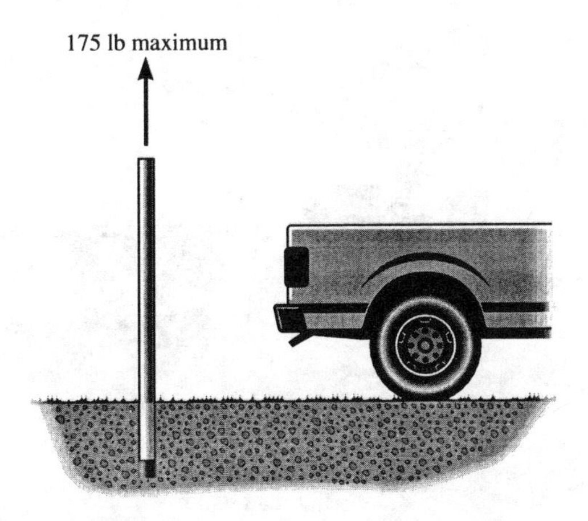

D–8. DESIGN OF A CART LIFT

A hand cart is used to move a load from one loading dock to another. Any dock will have a different elevation relative to the bed of a truck that backs up to it. It is necessary that the loading platform on the hand cart will bring the load resting on it up to the elevation of each truck bed as shown. The maximum elevation difference between the frame of the hand cart and a truck bed is 1 ft. Design a hand-operated mechanical system that will allow the load to be lifted this distance from the frame of the hand cart. Assume the operator can exert a (comfortable) force of 20 lb to make the lift, and that the maximum load, applied to the center of the loading platform, is 400 lb. Submit a scaled drawing of your design, and explain how it works based on a force analysis.

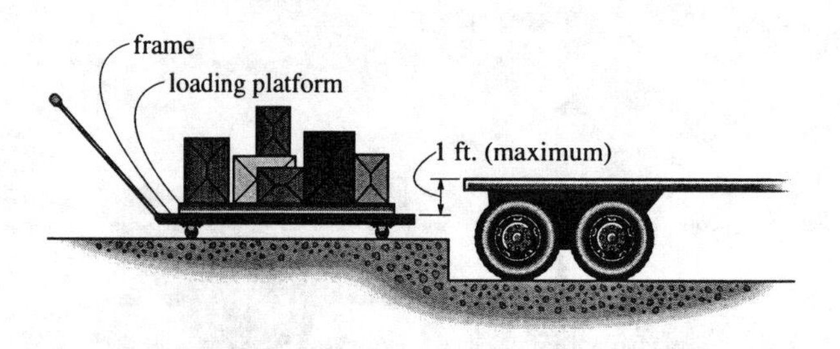

D–9. DESIGN OF A DUMPING DEVICE

Industrial waste is transported in dumpster carts to a centralized garbage bin. Each cart has the dimensions shown, and the maximum weight of a cart and its contents is anticipated to be 500 lb. Design a mechanical means of tipping the cart and allowing its contents to be dumped into the bin. Assume that a worker can exert a comfortable force of 20 lb on any handle used in the dumping operation. Submit a sketch of your design and explain how it works. Include a force analysis and discuss the safety and reliability of its use.

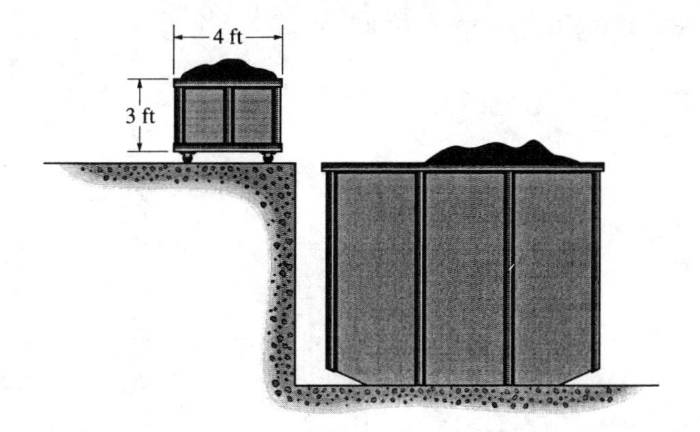

D–10. DESIGN OF A UTILITY POLE SUPPORTING STRUCTURE

An electric utility pole supports wires that exert a tension of 175 lb at the top of the pole in the direction shown. Although the pole is to be somewhat buried at its end *A*, it would be conservative to assume that this connection is a pin. Considering that the pole is located in a restricted space, devise a means of support that will hold the pole in a stable position. The support cannot be attached to the adjacent building or to a point on the sidewalk or roadway. Pedestrian clearance and safety should be considered in the design. Also, the economics of construction and its reliability should be addressed. Submit your design and discuss these issues. Also, provide the calculations used to determine the support reactions.

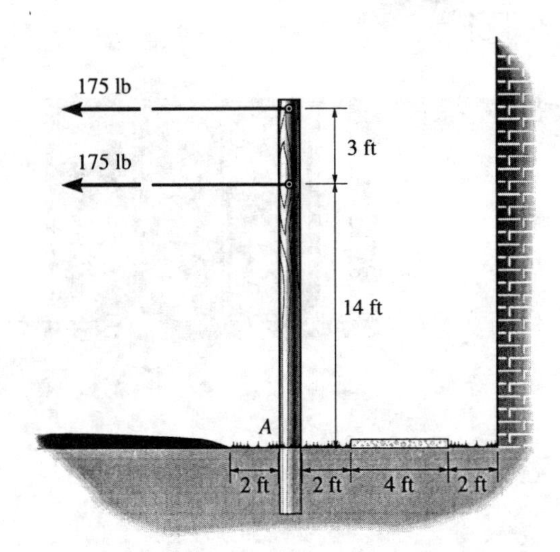

[To be assigned after coverage of Section 8.2]

D–11. DESIGN OF A ROPE-AND-PULLEY SYSTEM FOR PULLING A CRATE UP A RAMP

A large 300-kg packing crate is to be hoisted up the 25° ramp. The coefficient of static friction between the ramp and a crate is $\mu_s = 0.5$, and the coefficient of kinetic friction is $\mu_k = 0.4$. Using a system of ropes and pulleys, design a method that will allow a single worker to pull each crate up the ramp. Pulleys can be attached to any point on the wall *AB*. Assume the worker can exert a maximum (comfortable) pull of 200 N on a rope. Submit a drawing of your design and a force analysis to show how it operates. Estimate the material cost required for its construction. Assume rope costs $0.75/m and a pulley costs $1.80.

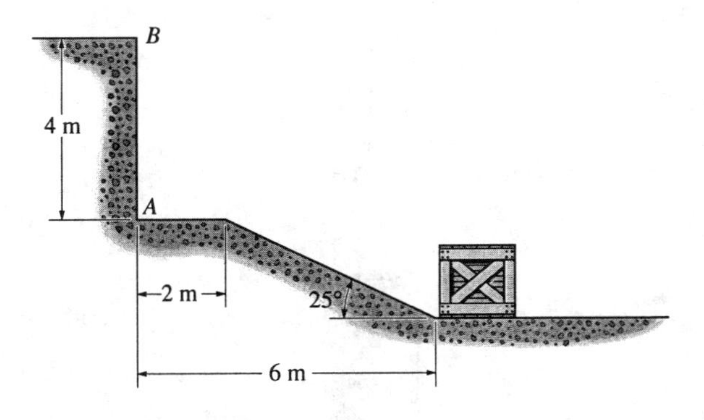

D–12. DESIGN OF A DEVICE FOR LIFTING STAINLESS-STEEL PIPES

Stainless-steel pipes are stacked vertically in a manufacturing plant and are to be moved by an overhead crane from one point to another. The pipes have inner diameters ranging from $100 \text{ mm} \leq d \leq 250 \text{ mm}$ and the maximum mass of any pipe is 500 kg. Design a device that can be connected to the hook and used to lift each pipe. The device should be made of structural steel and should be able to grip the pipe only from its inside surface, since the outside surface is required not to be scratched or damaged. Assume the smallest coefficient of static friction between the two steels is $\mu_s = 0.25$. Submit a scaled drawing of your device, along with a brief explanation of how it works based on a force analysis.

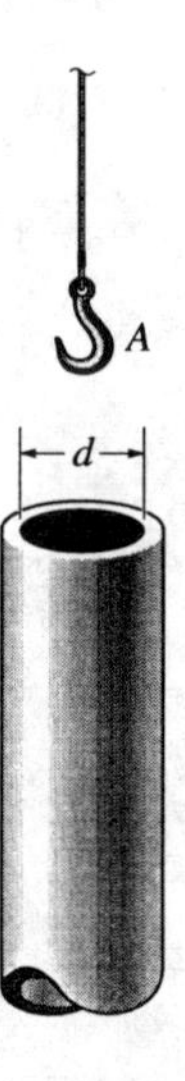

D–13. DESIGN OF A TOOL USED TO RELEASE A SUSPENDED LOAD

A 50-Mg load is suspended from the steel cable, and in an emergency it must be released from the cable so that it falls from the suspended position. Design a steel tool that can be connected to the cable at *A* and used to break the connection. The maximum (comfortable) pull that a worker can exert on the tool is 75 N. If needed, the coefficient of static friction of steel on steel is $\mu_s = 0.30$. Submit a scaled drawing of the tool, and a brief paragraph to explain how it works based on a force analysis. Include a discussion on the safety aspects of its use.

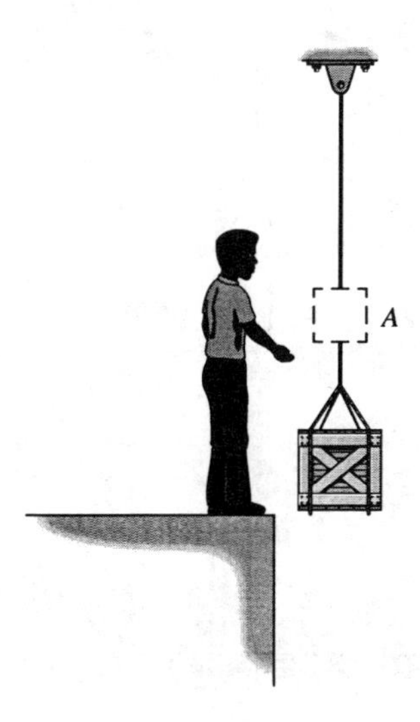

D–14. DESIGN OF A ROD-GRIPPING TOOL

Solid-steel 10-mm-diameter rods each have a mass of 50 kg and are to be individually transported to an assembly line by gripping the rod at one end and suspending it from a chain. Design a steel gripping tool to be used for this purpose, such that it is easy to connect and disconnect and will not damage the end of the rod. Submit a detailed drawing of your device, and include an explanation of how it works based on a force analysis. Discuss the safety aspects of its use. The coefficient of static friction between two steel surfaces is $\mu_s = 0.25$.

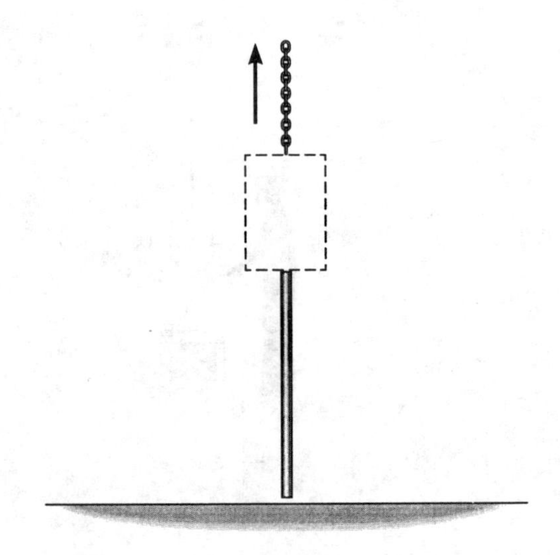

D–15. DESIGN OF A TOOL USED TO TURN PLASTIC PIPE

PVC plastic is often used for sewer pipe. If the outer diameter of any pipe ranges from 4 in. $\leq d \leq$ 8 in., design a tool that can be used by a worker in order to turn the pipe when it is subjected to a maximum anticipated ground resistance of 80 lb·ft. The device is to be made of steel and should be designed so that is does not cut into the pipe and leave any significant marks on its surface. Assume a worker can apply a maximum (comfortable) pull of 40 lb, and take the minimum coefficient of static friction between the PVC and the steel to be $\mu_s = 0.35$. Submit a scaled drawing of the device, and a brief paragraph to explain how it works based on a force analysis.

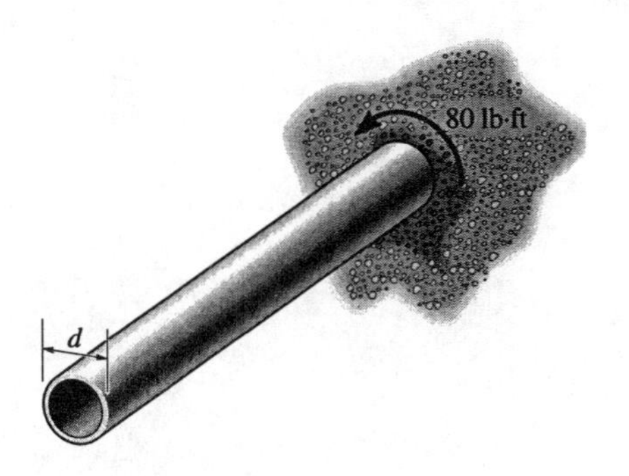